A Warm Place

by Susan Halko

Contents

Science Vocabulary

energy
Energy is something that can change things or do work.

The sun's **energy** warms and lights this beach in Florida.

evaporate
When water **evaporates,** it changes from a liquid to a gas.

Water **evaporates,** or turns into water vapor. Then it **condenses** and forms a cloud.

4

temperature

Temperature is a measure of how hot or cold something is.

In Florida, the **temperature** can be warm or cool in winter.

condense

When water vapor cools, it **condenses** and changes from a gas to a liquid.

Day and night is a **pattern.**

6

weather tool

A **weather tool** is something scientists use to measure the changing weather.

My Science Vocabulary

condense

energy

evaporate

pattern

temperature

weather tool

A thermometer is a **weather tool.**

Weather in Florida

Florida is mostly a warm place. Sometimes it's warm and sunny.

Sometimes there are clouds on a sunny day.

The sun's **energy** warms Earth.

energy

Energy is something that can change things or do work.

People and animals need the sun's energy to live. Most plants need the sun's energy to grow.

Living things need the sun's energy.

The sun's energy changes the weather.
It makes the air feel warm or hot.

It warms the water and land, too.

The Sun and Water

The sun **evaporates** water. The water turns into a gas called water vapor.

Water vapor condenses and turns into tiny drops of water. The drops form a cloud.

The sun's energy warms the water. The water evaporates and becomes water vapor.

evaporate

When water **evaporates,** it changes from a liquid to a gas.

Then water vapor **condenses**. It forms a cloud. Water falls from the cloud.

Water falls from the cloud. It's raining!

condense

When water vapor cools, it **condenses** and changes from a gas to a liquid.

The Changing Weather

The weather in Florida can change from day to day. It can change from hour to hour, too.

The Weather Changes from Day to Day

rainy	windy	sunny
Monday	Tuesday	Wednesday

The weather might be sunny in the morning.
Later in the day, storms might come.

Day, Night, Seasons

The weather changes as day turns into night. The air might feel cooler at night than it does during the day.

Day

The sun lights the sky during the day. The sky is dark at night. This is a **pattern.** It repeats over and over.

Night

pattern

A **pattern** is something that repeats over and over again.

The seasons follow a pattern, too.

Spring

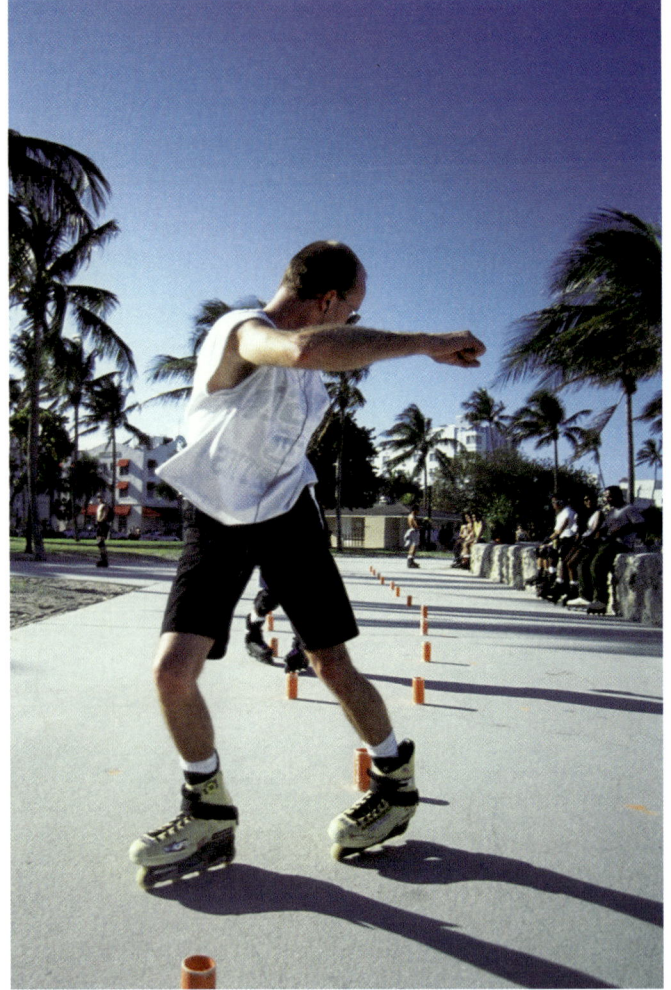

Summer

Florida has four seasons. They come in the same order year after year.

Fall

Winter

The weather changes from season to season. Winter is the coolest season. Some parts of Florida are cool. Some parts are warm.

Winter

People wear warm clothes on the beach in winter.

Winter turns into spring. The **temperature** rises.

Spring

Flowers bloom on orange trees in spring.

temperature

Temperature is a measure of how hot or cold something is.

Spring turns into summer. Summer is the warmest season. The air can feel hot and sticky.

In Florida, lightning is common in summer.

Summer turns into fall. The temperature goes down. Many storms happen. Winter is almost here!

Fall

In Florida, hurricanes happen in summer and fall.

Measuring the Weather

Scientists use **weather tools** to measure the weather. What do you think each of these tools measures?

Thermometer

Wind vane

Windsock

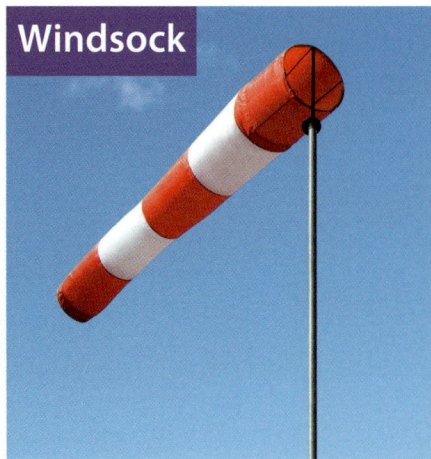

Rain gauge

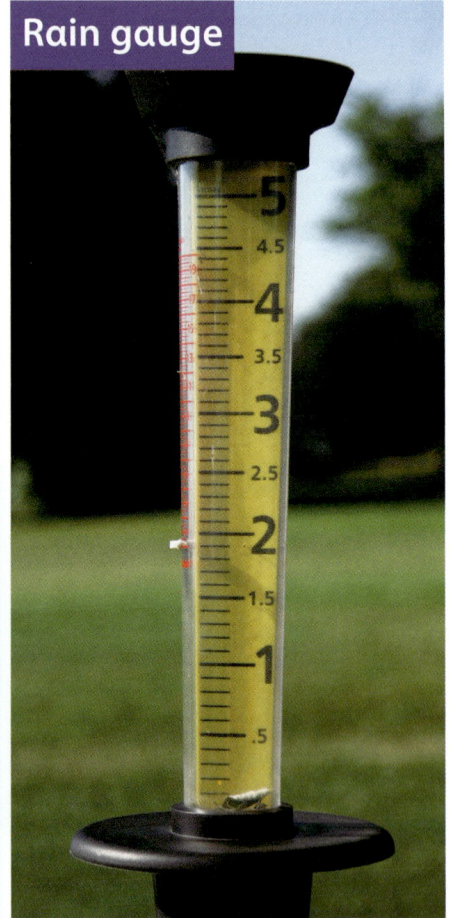

weather tool

A **weather tool** is something scientists use to measure the changing weather.

Florida gets tornadoes, hurricanes, and thunderstorms. It also gets lots of warm, sunny weather. The weather is always changing in Florida!

Conclusion

Florida is a warm place. But the weather changes. It changes every day. It changes every season. The sun's energy affects the weather.

Think About the Big Ideas

1. How does the sun affect the weather?
2. How does the weather change in Florida?
3. How is weather measured?

Share and Compare

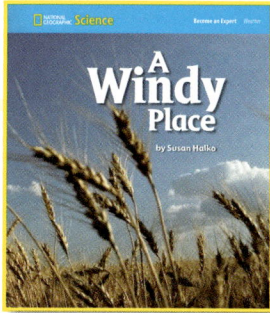

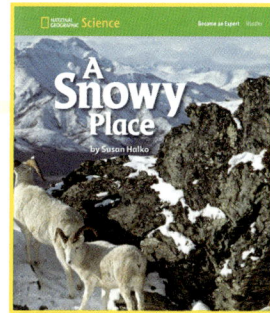

Turn and Talk

Compare the weather and seasons in different places. How are they alike? How are they different?

Read

Read your favorite caption to a classmate.

Write

Tell what winter is like in the place you read about. Share your writing with a classmate.

Draw

Draw two weather tools. Show what the tools measure. Share your drawings with a classmate.

Meet Tim Samaras

Tornadoes are dangerous storms. People can't observe them up close. So Tim Samaras invented a probe. It measures weather inside a tornado.

One day, information from the probe might help predict a tornado before it forms.

Index

Acknowledgments
Grateful acknowledgment is given to the authors, artists, photographers, museums, publishers, and agents for permission to reprint copyrighted material. Every effort has been made to secure the appropriate permission. If any omissions have been made or if corrections are required, please contact the Publisher.

Photographic Credits
Cover (bg) Tomasz Szymanski/iStockphoto; Cvr Flap (t), 4 (t), 10 S.Borisov/Shutterstock; Cvr Flap (c), 17, 28 Chuck Eckert/Alamy Images; Cvr Flap (b), 26 (r) Ryan Ruffatti/iStockphoto; Title (bg) Photos.com/Jupiterimages; 2-3 James P. Blair/National Geographic Image Collection; 5 (t), 22 Lisa F. Young/Shutterstock; 6 (t), 18 Joe Sohm/Visions of America, LLC/Alamy Images; 6 (b), 19 Artifacts Images/Digital Vision/Alamy Images; 7, 26 (l) nicholas belton/iStockphoto; 8-9 John Anderson/iStockphoto; 11 (t) Penrod Studios/Alamy Images, (b) moodboard/Corbis; 12-13 tbkmedia.de/Alamy Images; 16 (l) James Schwabel/Alamy Images, (c) Chad McDermott/Shutterstock, (r) slobo mitic/iStockphoto; 20 (l) Cheryl Casey/Shutterstock, (r) Danita Delimont/Alamy Images; 21 (l) Cheryl Casey/Shutterstock, (r) The Florida Times-Union, Carrie Rosema/AP Images; 23 Mauro Rodrigues/Shutterstock; 24 Mike Theiss/National Geographic Image Collection; 25 Claudio Lovo/Shutterstock; 26 (t) Duncan Walker/iStockphoto, (b) artpartner-images.com/Alamy Images; 27 Raul Touzon/National Geographic Image Collection; 30 Carsten Peter/National Geographic Image Collection; 31 2006 National Geographic; Inside Back Cover (bg) Gino's Premium Images/Alamy Images.

Illustrator Credits
4-5, 14-15 Greg Harris; 16 John Kurtz.

Neither the Publisher nor the authors shall be liable for any damage that may be caused or sustained or result from conducting any of the activities in this publication without specifically following instructions, undertaking the activities without proper supervision, or failing to comply with the cautions contained herein.

Program Authors
Kathy Cabe Trundle, Ph.D., Associate Professor of Early Childhood Science Education, The Ohio State University, Columbus, Ohio; Randy Bell, Ph.D., Associate Professor of Science Education, University of Virginia, Charlottesville, Virginia; Malcolm B. Butler, Ph.D., Associate Professor of Science Education, University of South Florida, St. Petersburg, Florida; Nell K. Duke, Ed.D., Co-Director of the Literacy Achievement Research Center and Professor of Teacher Education and Educational Psychology, Michigan State University, East Lansing, Michigan; Judith Sweeney Lederman, Ph.D., Director of Teacher Education and Associate Professor in the Department of Mathematics and Science Education, Illinois Institute of Technology, Chicago, Illinois; David W. Moore, Ph.D., Professor of Education, College of Teacher Education and Leadership, Arizona State University, Tempe, Arizona

The National Geographic Society
John M. Fahey, Jr., President & Chief Executive Officer
Gilbert M. Grosvenor, Chairman of the Board

National Geographic School Publishing
Hampton-Brown
www.NGSP.com

Printed in the USA.
RR Donnelley, Bedford, MA

ISBN: 978-0-7362-7565-1

11 12 13 14 15 16 17

10 9 8 7 6 5 4 3